Life and Adventures of Frances Namon Sorcho

The Only Woman Deep Sea Diver in the World

Captain Louis Sorcho

Originally Published in 1888

TABLE OF CONTENTS

THE BOTTOM OF THE SEA

PREFACE

HISTORY OF DEEP SEA DIVING

HOW MRS. SORCHO BECAME A DIVER

FEATS OF DIVERS

Captain Louis Sorcho

THE BOTTOM OF THE SEA

There's wondrous wealth to man unknown

At the bottom of the sea,

Where stately ships on Neptune's throne

Are rotting where the sun ne'er shone

In silent grandeur there alone,

At the bottom of the sea.

Mermaids dwell in caverns bright,

Jewels unknown flash their light,

And untold gold is hid from sight,

At the bottom of the sea.

Huge anchors lie buried in golden sand,

At the bottom of the sea.

Sunken junks from China's strand

'Mid old ship's splendid timbers stand,

Destruction dwells on every hand,

At the bottom of the sea.

Sightless fishes swim about

The bleaching bones, and in and out

The skulls of men, once brave, no doubt,

At the bottom of the sea.

The countless thousands sleeping there,

At the bottom of the sea.

The sailor, lover, maiden fair,

Who in the depths her jewels wear,

Now rest in peace without a care,

At the bottom of the sea.

Some are there well sewn in sail;

Ancient warriors clad in mail;

But one returns to tell the tale

of the bottom of the sea.

The diver, a mortal, like those that sleep

At the bottom of the sea.

An humble hero of the deep,

In sunken vessels' hulls doth creep

To wrest the golden treasure heap,

From the bottom of the sea.

In armor, helmet, shoes of lead,

She braves those awful depths of dread,

The living 'mongst the million dead,

At the bottom of the sea.

—By E. D. H.

Dedicated to FRANCES NAMON SORCHO

Captain Louis Sorcho

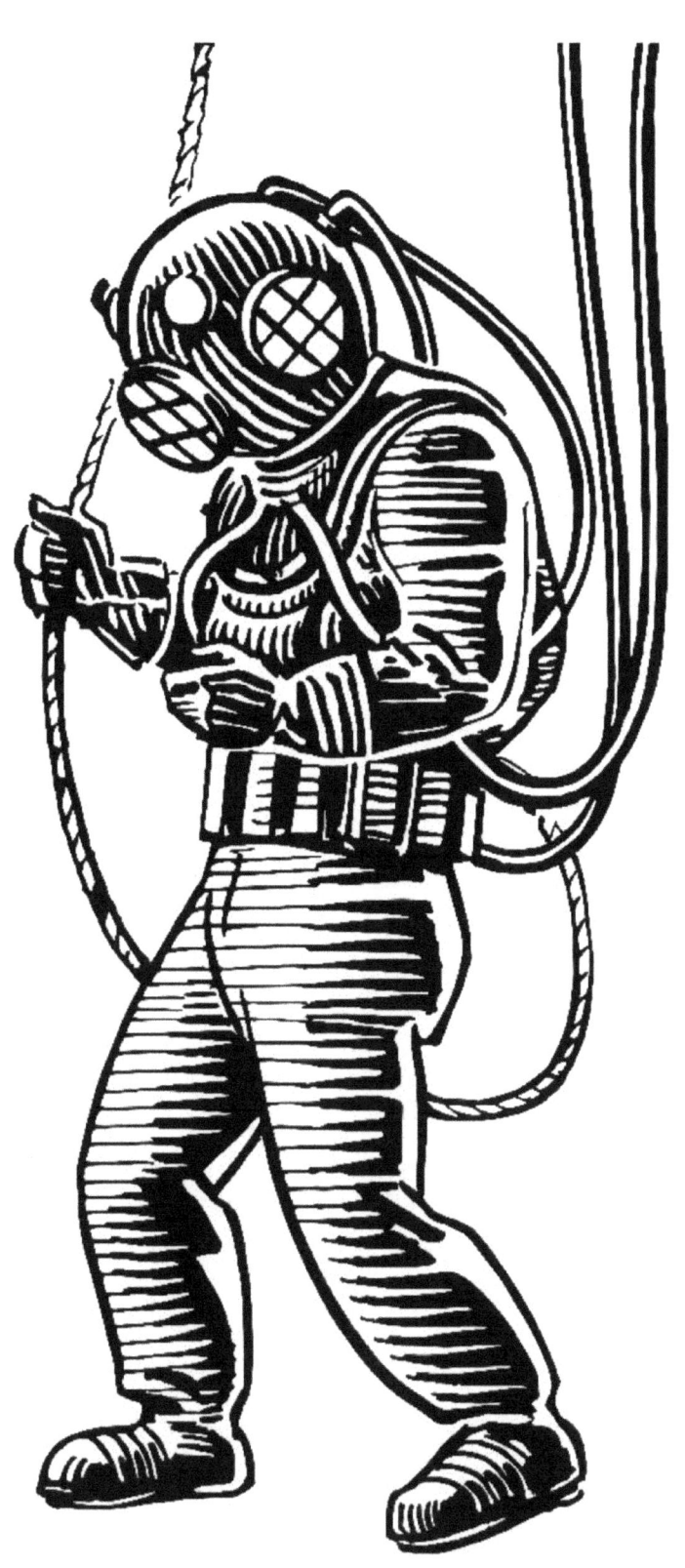

Captain Louis Sorcho

PREFACE

To the average individual unacquainted with the art of deep sea diving and the mysteries of the ocean away down beneath its surface, divers are sort of super-human creatures often read about but seldom seen. How they exist in the ocean's depths, the queer costume they are compelled to wear, the strange sensations they experience, the wonderful sights they see, the desperate risks they take, and the manner in which they work beneath the water, have, heretofore, all been a sealed volume to the general public.

In presenting this little book to our patrons, it is our object to enlighten them on these subjects, and give them some idea, at least, of the life of a diver.

Mrs. Sorcho is the only woman deep sea diver in the world, and is the only woman alive to-day who has ever donned a sub-marine armor and descended into the ocean's depths to work.

The example of intelligent daring is never lost on the world. The mastery of human beings over the material world is evident on every side, but too often are they themselves slaves to lesser things. With skill and courage, with caution and daring, with full knowledge of the danger, but with complete control over herself, this lady has accomplished what no other woman has ever dared to attempt. Fears, what are they? Coward thoughts. See Richard cowering in his tent. See infants crying in the dark. See here a woman, who has braved the thousand deaths that await the diver; who has calmly, yet courageously, ventured in the ocean's depths, with only the fishes and the thousands awaiting the day when the sea shall give up its dead for companions; kept herself in perfect control and invaded the mystic depths as a conqueror, mistress alike of element and herself. Heroism is a medieval thought, daring a classic record. To-day society languishes, passion chills, the spirit of adventure dies, the glory of arms is stilled by peace congresses, and human beings dwindle into a part of simple mechanism. Four-fifths of the dangers of life are as trifles, if met with courage, resolution and common sense.

Mrs. Sorcho is ready at any and all times to dive deeper and remain under water longer than any other female, or forfeit $10,000.

Our armor is of the latest improved pattern, with telephone, electric search-light, and many other up-to-date attachments invented and used exclusively by us. The scenes presented are exactly as they occur in the diver's life at the ocean's bottom, and the exhibition cannot fail to instruct and amuse both the old and young.

Trusting it may meet with your kind approval, we are,

Very sincerely yours,

CAPT. LOUIS SORCHO,

FRANCES NAMON SORCHO,

Deep Sea Divers.

Captain Louis Sorcho

HISTORY OF DEEP SEA DIVING

Just how far back the art of sub-marine diving dates, is a matter of conjecture, but until the invention of the present armor and helmet in 1839, work and exploration under water was, at best, imperfect, and could only be pursued in a very limited degree. The armor of to-day consists of a rubber and canvas suit, socks, trousers and shirt in one, a copper breastplate or collar, a copper helmet, iron-soled shoes, and a belt of leaden weights to sink the diver.

The helmet is made of tinned copper with three circular glasses, one in front and one on either side, with guards in front to protect them. The front eye-piece is made to unscrew and enable the diver to receive or give instructions without removing the helmet. One or more outlet valves are placed at the back or side of the helmet to allow the vitiated air to escape. These valves only open outwards by working against a spiral spring, so that no water can enter. The inlet valve is at the back of the helmet, and the air on entry is directed by three channels running along the top of the helmet to points above the eye-pieces, enabling the diver to always inhale fresh air, whilst condensation on the glasses is avoided. The helmet is secured to the breastplate below by a segmental screw-bayonet joint, securing attachment by one-eighth of a turn. The junction between the waterproof dress and the breastplate is made watertight by means of studs, brass plates, and wing-nuts. A life or signal line enables the diver to communicate with those above. The air-pipe is made of vulcanized india-rubber with galvanized iron wire imbedded.

The cost of a complete diving outfit ranges from $750 to $1,000. The weight of the armor and attachments worn by the diver is 246 pounds, divided as follows:—

Helmet and breastplate, 51 pounds; belt of lead weights, 122 pounds; rubber dress or suit, 19 pounds; iron soled shoes, 27 pounds each.

The greatest depth reached by any diver was 204 feet, at which depth there was a pressure of 88½ pounds per square inch on his body. The area exposed of the average diver in armor is 720 inches, which would have made the diver at that depth sustain a pressure of 66,960 pounds or over 33 tons.

The water pressure on the diver is as follows:—

20 *feet*	8½ *lbs.*
30 *feet*	12¾ *lbs.*
40 *feet*	17¼ *lbs.*
50 *feet*	21¾ *lbs.*
60 *feet*	26¼ *lbs.*
70 *feet*	30½ *lbs.*
80 *feet*	34¾ *lbs.*
90 *feet*	39 *lbs.*
100 *feet*	43½ *lbs.*
110 *feet*	47¾ *lbs.*
120 *feet*	52¼ *lbs.*
130 *feet*	56½ *lbs.*
140 *feet*	60¾ *lbs.*
150 *feet*	65¼ *lbs.*

The limit

160 *feet*	69¾ *lbs.*
170 *feet*	74 *lbs.*
180 *feet*	78 *lbs.*
190 *feet*	82¼ *lbs.*
204 *feet*	88½ *lbs.*

The air which sustains the diver's life below the surface is pumped from above by a powerful pump, which must be kept constantly at work while the diver is down. A stoppage of the pump a single instant, while the diver is in deep water, would result in his almost instant death from the pressure of the water outside. Only persons of perfect health and physique can pursue the calling of a diver. It would be suicidal for a human being not of perfect health and physique to attempt the feat.

Before a man attempts diving he should be examined by a physician or medical officer. Men coming under any of the following classifications should not, under any circumstances, attempt a dive. Men with short necks, full-blooded, and florid complexions. Men who suffer from headache, are slightly deaf, or have recently had a running from the ear. Men who have at any time spat or coughed up blood. Men who have been subject to palpitation of the heart. Men who are very pale, whose lips are more blue than red, who are subject to cold hands and feet, men who have, what is commonly known as, a poor circulation. Men who have blood-shot eyes and a high color on the cheeks, by the interlacement of numerous small blood-vessels, which are distinct. Men who are hard drinkers and have suffered from any severe disease, or who have had rheumatism or sun-stroke.

The dangers of diving are manifold, and so risky is the calling, that there are only a few divers in the United States. The cheapest of them command $10 a day for four or five hours work, and many of them get $50 and $60 for the same term of labor under water.

The greatest danger that besets the diver is not, as would doubtless be supposed, the monsters of the deep, such as sharks, etc., or of getting his air-hose entangled or fouled so as to cut off his air supply. It is the risk he runs every time he dives of rupturing a blood-vessel by the excessively compressed air he is compelled to breathe. Many divers have been hauled up dead in the armor from no apparent cause, when they had been plentifully supplied with air. In each case the rupture of a blood-vessel in the brain by the air pressure, had caused a fatal stroke of apoplexy. Divers have also died of fright in the armor. In one instance a diver at work in the hold of a sunken vessel got his air-hose so fouled, it was impossible to haul him up. Plenty of fresh air, however, was supplied to him, but he was held prisoner five hours, until another diver was procured to go down and free him. When he was hauled up he was a corpse. Fright had killed him. The diver is also subject to attacks by sharks, sword-fish, devil-fish and other voracious monsters of the ocean's depths. To defend himself against them, he carries a double-edged knife, as sharp as a razor, which screws into a watertight brass sheath, but is always ready for instant use. It is the diver's sole weapon of defense.

HOW MRS. SORCHO BECAME A DIVER

"As a girl in a quiet little home in Virginia, I little thought I would ever become a diver. In fact I didn't know what a real diver was.

"When I first saw the queer rig I shuddered, but now the grotesque costume is as natural to me as is my tea-gown, and perhaps I feel a little more at home in it.

"Only arms, limbs and a body well trained muscularly can walk about in shoes that weigh 27 pounds apiece, supporting an armor with copper helmet and breastplate, and leaden belt of weights which tip the scale-beam at 246 pounds. Therefore, the commencement of my education as a diver consisted of a year's training in a school of physical culture. When it was completed my muscles were as hard and springy as steel, and I felt no fear on the score of physical strength as I contemplated my first visit to the 'bottom of the sea.'

"My first dive was off the southern coast of Florida, not far from Clear Water Harbor. My husband was at the time engaged in the business of collecting rare shells and coral for several Northern Universities. I well remember how I felt when I first donned the armor. Fear and curiosity were so closely blended that I hardly know which I felt the most of. At any rate, my husband was waiting, and almost before I realized it the queer canvas armor had been adjusted and the breastplate had been slipped over my head. A thick pad or collar had been put on my shoulders to take the weight off the breastplate and helmet, which alone weigh 56 pounds; but even then the plate felt quite heavy, and as the metal gaskets were being screwed down with thumb-nuts and a wrench, I felt as if I were being screwed up in my coffin. But there was little time for such gruesome reflections, and a stout leather belt holding the sub-marine knife was next girded about my waist.

"This knife, a double-edged affair, sharp as a razor, screws into a watertight brass scabbard. It is the diver's only weapon, and with it he must protect himself against sharks and other sub-marine monsters. The shoes come next. How heavy and awkward they looked, with their soles of cast-iron two inches thick, and how clumsy they felt when I tried to walk in them for the first time!

"The life-line—that all-important half-inch manilla rope—was then knotted about my waist, and the belt of leaden weights was strapped about me under the arms, and I was told to step over the railing of the boat on to the short ladder that had been suspended over her stern. I did so, mechanically I fear, and when I had managed to get down a few steps, the helmet was slipped over my head and by a deft turn locked.

"The queer headpiece was much larger than my head, and admitted of considerable freedom of movement inside it.

"'Now recollect,' said my husband, 'if you want to come up quick in case anything happens, give one jerk on the life-line. If you want more air give two jerks, or less air three jerks.'

"I expected to shoot to the bottom like a lump of lead, owing to all the weight I had on me, but I sank gradually instead, so buoyant was the inflated armor. I was on the bottom with five fathoms of water over my head almost before I realized it.

"I felt a sensation of pressure on the chest, and in my ears and head, which was quite painful. The first thing that I noticed, was a boiling of the water about me for which I was unable to account, until I happened to think of the foul air escaping through the valve in the back of the helmet.

"I found, also, to my surprise, that I could see quite well some distance about me, and observed a number of little fishes, which finally swam quite close to me and appeared to gaze in the glass front of the helmet with their little bead-like eyes, as though wondering what sort of a fish I was. I felt strangely light and buoyant, and found that with the slightest upward movement I would shoot surfaceward several feet. The armor also felt so stiff and hard that I could scarcely move in it.

"The next time I went down was not on a pleasure trip, but to work, and for several weeks my husband and I took turns diving for shells and curios. We finally completed our contract.

"Recovering a dead body is the task a diver dislikes more than any other kind, and although I have recovered quite a number, the work is yet horrible to me.

"The first dead body I ever brought to the surface was that of a man who was supposed to have been murdered and thrown into a lake near Atlanta, Ga. I searched the entire bottom of the lake, and finally in a deep hole found the body.

"It was shockingly mutilated and disfigured, and was almost unrecognizable, but we never found out whether the man had been murdered or not.

"When I came to the surface with that bloated, disfigured corpse, strong men were made sick and turned away, and to tell the truth I felt a little squeamish myself; but it was a matter of business, not sentiment, with me, so I doffed the armor and pocketed the reward that had been offered.

"The exploding of sub-marine torpedoes is dangerous work, and you can take my word for it that one does not feel very comfortable groping about with five or six pounds of dynamite in her hand, not knowing what minute it may take a notion to go off and blow her into kingdom come. Diving is fascinating, but it is dangerous, and there are very few women who would care to engage in it even if they had the nerve."

Captain Louis Sorcho

FEATS OF DIVERS

Millions of dollars worth of property has been recovered from the ocean's depths by divers. One of the greatest achievements in this line was by the famous English diver Lambert, who recovered vast treasure from the Alfonso XII, a Spanish mail steamer belonging to the Lopez Line, which sank off Point Gando, Grand Canary, in 26½ fathoms of water. The salvage party was dispatched by the underwriters in May, 1885, the vessel having £100,000 in specie on board. For nearly six months the operations were persevered in, before the divers could reach the treasure-room beneath the three decks. Two divers lost their lives in the vain attempt, the pressure of water being fatal. Mr. Gorman recovered £90,000 from the wreck, and got £4,500 for doing it.

One of the most difficult operations ever performed by a diver, was the recovering of the treasure sunk in the steamship Malabar off Galle. On this occasion the large iron plates, half an inch thick, had to be cut away from the mail room, and then the diver had to work through nine feet of sand. The whole of the specie on board this vessel—upward of $1,500,000—was saved, as much as $80,000 having been got out in one day.

It is an interesting fact that from time to time expeditions have been fitted out, and companies formed, with the sole intention of searching for buried treasure beneath the sea. Again and again have expeditions left New York and San Francisco in the certainty of recovering tons of bullion sunk off the Brazilian coast, or lying undisturbed in the mud of the Rio de la Plata.

At the end of 1885, the large steamer Indus, belonging to the P. & O. Co., sank off Trincomalee, having on board a very valuable East India cargo, together with a large amount of specie. This was another case of a fortune found in the sea, for a very large amount of treasure was recovered.

Another wreck, from which a large sum of gold coin and bullion was recovered by divers, was that of the French ship L'Orient. She is stated to have had on board specie of the value of no less than $3,000,000, besides other treasure.

A parallel case to L'Orient is that of the Lutine, a warship of thirty-two guns, wrecked off the coast of Holland. This vessel sailed from the Yarmouth Roads, with an immense quantity of treasure for the Texel. In the course of the day it came on to blow a heavy gale; the vessel was lost and went to pieces. Salving operations by divers, during eighteen months, resulted in the recovery of $400,000 in specie.

Another remarkable case of recovery of specie is recorded, when sixty-two chests of dollars, amounting to the value of about $350,000, were recovered from the Abergavenny, sunk some years previously at Weymouth, England.

A very notable case—not only for the amount of treasure on board, but also for the big "windfall" for the salvors—is that of the Thetis, a British frigate, wrecked off the coast of Brazil, with $800,000 in bullion on board. The hull went to pieces, leaving the treasure at the bottom in five or six fathoms of water. The admiral of the Brazil station, and the captains and crews of four sloops-of-war, were engaged for eighteen months with divers in recovering the treasure. The service was attended with great skill, labor and danger, and four divers' lives were lost.

A remarkable case of money having been recovered deserves a passing notice. It was that of the finding of 3,800 sovereigns under a pier at Melbourne, part of 5,000 missing from the steamer Iberia.

Some Danish speculators are reaping a harvest of golden grain from the depths of the sea which washes the coast of Jutland. Some years ago, the British steamship Helen, laden with copper, foundered. All her cargo has been recovered. The steamer Westdale, laden with 2,000 tons of iron, went down off the Danish coast in 1888. Nearly the whole cargo, her machinery, and a great part of her fittings, have been saved by Jutland divers.

Dredging operations carried on at Santander, Spain, resulted in the discovery of the well-preserved wreck of a warship of the fifteenth or sixteenth century. She must have been in her present position for four hundred years, and was partly covered by a deposit of sand and mud. Divers brought up guns which bore the united arms of Castile and Aragon, the scroll of Isabella, or the crown and initial of Ferdinand. The ship was probably employed as a transport, and inasmuch as some of the arms are of French and Italian make, it is supposed she formed part of the fortunate expedition against Naples under Gonzalo de Cordoba.

Captain Louis Sorcho

Captain Louis Sorcho

www.ingramcontent.com/pod-product-compliance
Lightning Source LLC
Chambersburg PA
CBHW040109120526
44589CB00040B/2835